INSTITUTE OF PETROLEUM

Pressure Vessel Examination

INSTITUTE OF PETROLEUM

Pressure Vessel Examination

Being Part 12 of the
Institute of Petroleum
Model Code of Safe Practice in the
Petroleum Industry

2nd Edition

March 1993

Published on behalf of

THE INSTITUTE OF PETROLEUM, LONDON

by

John Wiley & Sons

Chichester • New York • Brisbane • Toronto • Singapore

Library of Congress Cataloging-in-Publication Data

Institute of Petroleum (Great Britain)
      Pressure vessel examination / the Institute of Petroleum. — 2nd
ed.
      p.   cm. — (Model code of safe practice in the petroleum
industry / Institute of Petroleum : pt. 12)
  'Published on behalf of the Institute of Petroleum, London.'
  'March 1993.'
  ISBN 0 471 93936 6 (paper)
    1. Pressure vessels — Testing. 2. Pressure vessels — Inspection.
3. Pressure vessels — Safety measures. I. Title. II. Series: Institute of
Petroleum (Great Britain). Model code of safe practice in the
petroleum industry; pt. 12.
TS283.I535 1993   92-45609
681'.76041'0287—dc20   CIP

British Library Cataloguing in Publication Data

A catalogue record for this book is available
from the British Library.

ISBN 0 471 93936 6

Printed and bound in Great Britain.

CONTENTS

FOREWORD

This Code is Part 12 of the IP Model Code of Safe Practice in the Petroleum Industry. Its purpose is to provide a guide to safe practices in the in-service examination and test of pressure vessels used in the petroleum and chemical industries.

The Code gives general requirements regarding the provision and maintenance of adequate documentation, in-service examination, the control of modifications and repairs, examination frequency, protective devices and testing of pressure vessels.

In many countries there exist statutory requirements, both local and national, pertaining to the in-service examination of pressure vessels and, where this is so, this Code should be regarded as being complementary to such requirements.

Although it is believed that the adoption of this Code will help to reduce the risk of accident, the Institute cannot accept any responsibility of whatsoever kind, for damage or alleged damage arising or otherwise occurring in or about premises or areas to which this Code has been applied.

For the purpose of this Code, certain definitions, which are given in Chapter 1, Section 1.3, apply irrespective of any other meaning the words may have in other connections.

ACKNOWLEDGEMENTS

In preparing and revising this Code, reference has been made to the existing Codes and Standards of most of the principal petroleum, petro-chemical and chemical companies operating in the United Kingdom, who have given permission to use appropriate material in the preparation of this Code.

The Institute of Petroleum wishes to acknowledge the helpful co-operation of these companies and also that of the Oil Companies Materials Association who materially assisted in the preparation of the original of the Code, first published in 1976; further acknowledgement is due to the Engineering Equipment and Materials Users Association (which now incorporates OCMA) and the Committee of User Inspectorates for assistance with revision and updating of the Code prior to re-issue in 1993.

1

GENERAL

1.1 INTRODUCTION

This Code was first published in 1976. Since then there have been considerable improvements to the administration of inspection bodies — resulting from more formal documentation of procedures, authorities and responsibilities and from improved record keeping. There have also been significant advances in inspection technology, on-line monitoring and non-destructive testing — involving such techniques as X-ray and thermal imaging, computerized records and data storage and analysis, the application of fracture mechanics, flash photography, on-line testing of relief devices, portable spectrometers, and in-situ replication examination. In addition the latest design and construction codes together with appropriate quality systems have resulted in a higher level of quality assurance in the manufacturing of new equipment.

The Code has thus been revised to reflect the changes of the past 16 years and their effect upon the safe and efficient operation of pressure plant. Also, whilst the Code remains an international document, the opportunity has been taken to effect some re-arrangement of the text and adjustment of the terminology to align it with the (UK) Pressure Systems and Transportable Gas Containers Regulations 1989, hereinafter referred to as PSRs.

1.2 SCOPE

This Code is for the guidance of users of equipment operating over a wide range of pressure and temperature, together with their associated protective safety devices, but excluding parts of prime movers, compressors, pumps and like machinery, pipes and coils, portable or transportable vessels.

It defines a procedure for the identification, classification and monitoring of equipment to ensure that its fitness for purpose is maintained.

The objectives of the Code are:

(a) to provide guidance on the requirements of a structured examination system to ensure inspection activities are prudently managed,

(b) to provide basic examination and test procedures,

(c) for the UK only, to assist users of pressurized equipment to implement PSRs,

(d) for the UK only, to form a basis for that part of the written scheme applicable to pressure vessels.

The advice given is based on petroleum and chemical industry practices which, through operating experience, have proved to be both necessary and beneficial for safe and economic operation as well as protection of environment.

This approach consists primarily of regular, scheduled examinations and reviews by Competent Persons who may be employed by the owner/operator Inspection Services or by an External Inspecting Authority. The Code does not include rules for the design of pressure vessels, and it is assumed that the equipment subject to examination has been designed and constructed to recognized Codes and Standards appropriate to its intended service conditions.

It may be appropriate to treat vessels subject to negative pressure in a similar manner to pressure vessels as defined in Section 1.3.

1.3 DEFINITIONS

(a) Competent Person

That person or body delegated or authorized by the User to be responsible for drawing up or approving Schemes of Examination and for examination of pressure vessels within the pressure system.

(In the UK the Competent Person may be Company personnel or may be an external body specializing in the inspection of pressure vessels. Where applicable, reference should be made to the Approved Code of Practice relating to the PSRs for detailed attributes of a Competent Person).

(b) Design Authority

The Design Authority may be a Vessel Design group responsible to the User, an authorized Design Contractor, an independent Design Consultant, the Engineering Authority or the Competent Person, as designated by the User.

In those countries where design, construction and operation of pressure vessels is controlled or regulated by State-recognized Agencies, those Agencies will be regarded as the Design Authority.

(c) Engineering Authority

The person or persons delegated and authorized by the User to be responsible for the upkeep and maintenance of pressure vessels utilized within the pressure system.

(d) Examination

An examination means a careful and critical scrutiny of pressurized equipment by a Competent Person to assess:

(i)   its actual condition, and
(ii) the period for which it may be safely used prior to the next examination.

(e) Modification

Any change in the vessel, temporary or permanent, which may affect the safety and integrity of the system.

This also includes any changes in process materials, additives, services, operating conditions or operating procedures, and non-routine programmes which fall outside normal operating instructions.

(f) Pressure Vessel

A closed vessel consisting of one or more independent chambers, any or each of which may be subject to an internal pressure greater than 0.5 barg, or as defined by National Legislation.

Excluded are vessels subject to pressure solely by static head and storage tanks designed and constructed to Standards or Codes such as BS 2654, BS 2594.

(g) Protective Device

A device designed to protect the pressure system against system failure and certain devices designed to give warning that system failure might occur, safety valves, bursting discs or combination of both.

(h) Safety Valve

A valve which automatically, without assistance of any energy other than that of the fluid concerned, discharges a sufficient quality of fluid so as to prevent a predetermined safe pressure being exceeded, and which is designed to re-close and prevent the further flow of fluid after normal pressure conditions of service have been restored.

(i) Bursting Disc

A Bursting disc is a non-re-closing pressure relief device actuated by differential pressure and designed to function by the bursting or venting of a pressure-containing and pressure-sensitive component.

(j) Scheme of Examination

A written scheme, defining the extent, nature and frequency of examinations. (in the UK, reference should be made to Regulation 8 of PSRs.)

(k) User

The person who has control of the operation of a pressure system of which the pressure vessel is a part.

2

REGISTRATION AND CLASSIFICATION

2.1 REGISTRATION

All pressure vessels and protective devices within the scope of this Code shall be allocated a unique Identification Number. This number shall be stamped or otherwise clearly marked on the vessel or equipment and, in addition, for pressure vessels, this number should be painted on the vessel, its lagging or cladding, in an easily visible place. In the UK such marking shall be in accordance with Schedule 4 of the PSRs.

If the vessel or protective device is transferred to a different duty and/or location and the Identification Number requires to be changed then the original Identification Number shall be removed and/or obliterated and a new number allocated and marked on the vessel or protective device in a similar location to the previous one. Alternatively, the Identification Numbering system adopted may be such that the number allocated to an item remains constant despite a change of duty. In this case, the documentation relating to the item must clearly record and identify the change of duty.

Each item which is allocated an Identification Number shall have a Records File created and maintained. This file may be either paper or secure electronic system. No pressure vessel or protective device shall be registered unless satisfactory documents and drawings are available regarding their design and manufacture and the equipment is suitable for its intended service duties.

The Records File should include where possible:

Plant or System Identification Number.
Order Number.
Drawing References.
Specifications.
Materials lists.
Mill Test Certificates for Materials.
A Facsimile or rubbing of the Code Stamp applied by the Manufacturer or External Inspection Authority.
Examination Reports during manufacture.
Final Acceptance and Test Certificates.
Safe Operating Limits or reference thereto (as defined in the PSRs).

It should also contain information on the design and operating conditions, the nature of the service duty, notes of any unusual features or causes for possible deterioration to which the item may be subjected and the date of entering service.

The file shall also, where applicable contain the Scheme of Examinations.

Where available retiring limit criteria should be included together with technical notes on the derivation of these criteria. If a vessel is subsequently down-graded to lower operating conditions the criteria shall be revised. A copy of all subsequent Examination Reports and Technical Data Reports should be kept in the File, together with details of modifications and/or repairs, changes of duty and the necessary authorization for any such changes.

Where such details or papers cannot be kept in the Records File, reference should be made to the location where this information is preserved.

For Protective Devices, the File should contain a complete specification of duty, the supplier details of the style and type of device, materials of construction, special features which may be incorporated in the device and information (including diagrams or drawings or reference to location of master copies) of the function and operation of the device. To this should be added copies of all reports of subsequent examinations and tests.

Where, in the case of safety valves attached to or protecting a pressure vessel the conditions of service of that pressure vessel are materially altered then the set pressure and discharge capacity of that safety valve shall be recalculated and the calculations retained in the Records File or reference made to where this information is preserved.

2.2 CLASSIFICATION

All registered equipment shall be classified as indicated below.

2.2.1 Classification A

This Class shall include all vessels and their Protective Devices which are subject to periodic examination in accordance with National or Regional Rules and Regulations. The extent of this classification will vary according to the country or state in which the equipment is to be used.

UK Requirements — Classification A applies to equipment that falls within the scope of Regulation 8 of PSRs. In other countries use should be made of statutes which apply in that country.

Where National or Regional Rules and Regulations permit Pressure Vessels or Protective Devices in Class A should be further allocated to:

Grade 0, Grade 1, Grade 2 or Grade 3 in accordance with the principles enunciated in Chapter 3. Equipment may be regraded following a period of service subject to the conditions laid down in Chapter 3.

2.2.2 Classification B

This class shall include all vessels and their Protective Devices not included in Classification A — i.e. Class B includes all vessels and their Protective Devices for which there is no legal obligation for periodic examinations.

In the UK, legislation is primarily directed towards ensuring that equipment containing significant stored energy from compressible fluids is periodically examined. Many items of equipment which contain toxic or flammable liquids are effectively exempt from specific periodic examination requirements. Nevertheless, periodic examination for this type of equipment has been found beneficial for safety, environmental and economic reasons.

This code suggests that the User of Class B equipment should therefore apply the same inspection philosophies and guidance on intervals as those described in Chapter 3.

2.3 CHANGE OF DUTY

Where it is proposed to transfer a vessel of Protective Device to a different duty, the Classification of the equipment shall be subject to a Review, particularly the Grading Allocation, which shall be reassessed based on knowledge of the condition of the vessel, its new duty and the grading of the equipment which it replaces.

Where a change of operating conditions is proposed (including increased throughput) careful consideration should be given to the likely or possible effects on the life or safe operation of the equipment and the Grading Allocation chosen or modified accordingly.

In both the above cases, the User should obtain approval to make the proposed change from the Competent Person who should consider the effects of changes in pressure, temperature, throughput, additional loadings, liability to corrosion, stress corrosion, corrosion, fatigue, creep, hydrogen attack, high temperature sulphur attack, etc., setting and capacity of protective devices.

Significant changes may require a thorough reappraisal of the original design criteria by a Design Authority.

3

EXAMINATION INTERVALS AND GRADING

3.1 GENERAL

The selection of an examination interval and the Grading Allocation is a matter for experienced judgement of personnel thoroughly familiar with all aspects of the equipment and its duty and should be approved by the Competent Person.

This code suggests two concepts which interrelate and affect decisions regarding examination intervals:
(a) the allocation of Grades,
(b sampling examination procedures, all of which are further discussed and explained below.

3.2 PRINCIPLES OF EXAMINATION FREQUENCY

Where National or Regional Rules and Regulations permit pressure vessels and protective devices should be allocated to a Grade 0, 1, 2 or 3 which indicates the maximum interval which may elapse between major examinations. Each item should:

(a) receive a precommissioning examination before entering service for the first time;
(b) initially be given a Grade 0 and be given its first thorough examination following a comparatively short period of service;
(c) subsequently, on the basis of a knowledge of service conditions and the condition of the equipment following the first thorough examination either be retained in Grade 0 or be given a Grade of 1 or 2;
(d) subsequently, on the basis of further knowledge gained of service conditions and the condition of

the equipment following the first and second thorough examination, be given an examination Grade of 1, 2 or 3 as appropriate to its needs.

When approaching the Design or predicted Remanent Life of the vessel it may be necessary to revert back to a more stringent examination grade.

Certain special case exceptions are permitted in Sections 3.3.5, 3.3.6 and 4.3 of this Code, and guidance for intervals for UK steam and air equipment is given in the Approved Code of Practice to PSRs, under Regulation 8.

Table 1 states the maximum intervals which may be allowed to elapse between examination. One postponement beyond the due date for examination may be granted subject to each case being agreed in writing by the user and the Competent Person. Under Reg 9 (7) of PSRs in the UK such postponements shall be notified to the Enforcing Authority.

Table 1. Examination frequency

| Equipment | Recommended maximum examination period (months) | | | |
	Grade 0	Grade 1	Grade 2	Grade 3
Process pressure vessels and heat exchangers	36	48	84	144
Pressure storage vessels	60	72	108	144
Protective devices	24	36	72	—

The intervals recommended in Table 1 are maxima. Intervals less than those given in table may be stipulated if more appropriate to the conditions. It should be noted the maximum recommended examination period for vessels is 144 months.

3.3 EXAMINATION GRADING ALLOCATION

3.3.1 Examination Grade 0

All equipment shall be deemed to be Grade 0 and shall remain in this grade until a first examination is carried out, except as permitted in Sections 3.3.5, 3.3.6 and 4.3.

In addition, equipment should be allocated to this Grade when the conditions of service are such that:
(a) deterioration in whole or in part is reasonably foreseeable at a relatively rapid rate, but consistent with the examination interval of this grade; or
(b) there is little evidence or knowledge of operational effects on which to predict behaviour in service.

3.3.2 Examination Grades 1 and 2

Equipment should be allocated to one of these Grades when the conditions of service are such that:
(a) deterioration in whole or in part has been shown to be at a reasonable and predictable rate consistent with the increased examination interval given for the item under the Grade selected, or
(b) evidence or knowledge of actual behaviour in service is sufficiently reliable to justify the examination interval permitted by the Grade selected, or
(c) there are established and reliable means of assessing operational effects and/or deterioration of the equipment that may give rise to danger.

3.3.3 Examination Grade 3

Equipment may be allocated to this Grade when the item has successfully concluded a period of service in Grade 2 and service conditions are such that:
(a) deterioration in whole or in part has been shown to be at a low predictable rate consistent with the increased examination interval given for the item in this Grade, or
(b) evidence and knowledge of actual service conditions are sufficiently accurate and reliable that an increased examination interval is justified.
Other factors to be considered in the choice of Grading are detailed in Sections 5.2 and 6.2.

3.3.4 Grading Transfers on the Basis of Examination

On the basis of the first examination, an item of equipment may be transferred to Grade 1 or Grade 2 provided that the examination has shown that the conditions for this new Grading have been met.

On the basis of subsequent examinations, items may be progressively transferred to Grade 2 or Grade 3 provided that the examination carried out has verified that the conditions for the new Grade frequency have been met. Conversely, and also on the basis of an examination, any item of equipment in Grade 2 or 3 shall be transferred back into a lower grade if the results of the examination show that conditions for the higher Grade are not being met.

3.3.5 Sample Examination of Vessels

Where a group of vessels are *substantially the same* as regards geometry, design, construction and conditions of service such that they may reasonably be expected to be subject to similar deterioration and that Grade 1 or above is appropriate for each, the following sample limits shall apply:

Number of items in group	Number of items in sample
2, 3 or 4	1
5, 6, 7 or 8	2
9 or more	3

Sample examinations of vessels in the group may commence at the first examination following commissioning and may continue to be applied at subsequent examinations subject always to the proviso that each individual vessel shall be given an examination within the maximum period allowed for Grade 3 (144 months).

In the event that findings of the sample examinations are unsatisfactory, it will be necessary to extend the scope of the sample to include other vessels from that group until sufficient information acquired relating to the group.

This section shall **not** be applied to **Protective Devices**.

3.3.6 New Vessels in Known Service Conditions

Exceptions to the normal Grading procedures may be applied in cases where a vessel:
(a) will perform a duty similar to that of an existing vessel and
(b) is substantially the same as the existing vessel as regards geometry, design, construction and conditions of service.
In such cases the new vessel may be given the same Grade as the vessel with which it is being compared.

3.3.7 Grading Review

Equipment shall be subject to a Grading Review when:
(a) significant changes have taken place in the

conditions of service of any registered items in any Grading allocation which would affect its deterioration in whole or part, or

(b) following an abnormal incident which has or could have affected the safety of operation of the equipment, or

(c) 'on-stream' inspection from corrosion probes, metal content of process streams, pH levels, etc. has indicated that there has been a significant change in the condition of the vessel that could warrant a change in the Grading Allocation, or

(d) the equipment approaches its intended design life or when it is proposed to extend the service life of the equipment beyond its original design life.

3.4 EXAMINATION INTERVALS AND GRADING FOR PROTECTIVE DEVICES

It is of paramount importance that Protective Devices give adequate protection to the equipment on which they are installed, and experience suggests that the periods between examinations for conventional pressure-actuated devices should be less than may be permitted for pressure vessels. These intervals must be based on knowledge of service experience and conditions found during shop overhaul and test.

Table 1 puts a limitation on the maximum period that any safety device may remain in service at 72 months. It must be emphasized that the periods quoted in any of Grading Allocations 0, 1 or 2 can be permitted only on the basis of satisfactory service experience on a particular duty, and evidence of a previous satisfactory performance shall be recorded in the Records File. Unsatisfactory performance shall be cause for review of the Grading Allocation and a reduction in the interval permitted to elapse up to the next examination.

In general the intervals between examinations on protective devices do not exceed the intervals between examinations of the pressure vessels protected.

In addition to conventional devices such as safety valves and bursting discs, certain specialized types of instruments may be characterized as protective devices if they have the primary function of protecting a vessel from the effects of pressure or temperature which the vessel was not designed to withstand. An example would be a temperature-actuated device to cut off the supply of energy or fuel during a temperature run-away in an exothermic reaction.

Such devices shall be registered items and may require in-situ testing at intervals much more frequent than Grade 0 to monitor their continued ability to safeguard the equipment they are designed to protect.

4

EXAMINATION PRINCIPLES

4.1 INTENT

The reasons for requiring examination of Pressure Vessels and Protective Devices may be summarized as follows:

> To ensure that equipment remains in a satisfactory condition for continued operation consistent with the prime requirements of safety, compliance with statutory regulations and economic operation until the next examination.

Chapter 3 gives guidance on the inspection frequency to be applied, which is allied to the Examination Grading Allocation. The type of examination at these intervals is outlined below.

4.2 TYPES OF EXAMINATION

4.2.1 Precommissioning Examination

The examination of each item of equipment carried out before commissioning is to ensure that:
(a) the Competent Person is satisfied that the specified examination and tests have been completed during manufacture and that the required documentation including reference to significant imperfections is in place, and
(b) a record is made of the new condition of the equipment
   — so as to act as a sound basis for judgement of deterioration when compared with the results of subsequent examinations,

— to confirm that no damage has occurred since examination during manufacture.

Such examinations may be undertaken by nominated inspection authorities prior to commissioning.

4.2.2 First Examination

The first examination of each item of equipment is carried out to ensure that:
(a) any apparent errors of design or materials are identified,
(b) selected components in the item are examined and measured (or otherwise inspected) and that such measurements (or inspection results) are recorded to form the basis of calculations or assessments of corrosion or wastage patterns for future safe life predictions.

4.2.3 Subsequent Examinations

Subsequent examinations are essentially follow-up examinations and are to ensure that the vessel shell and all its essential components and fitting are examined and measured and that the results of such measurements and examinations are compared with previous results, to enable assessment of corrosion or wastage patterns or other forms of possible defects to be made, and the future safe life remaining to be reliably predicted.

4.3 SPECIAL CASES

Design and Construction

It is recognized that special cases may be identified

which, due to the form of design, construction and/or the conditions of service, the principles enunciated in this Code are not readily applicable. This may apply, for instance, to certain items in cryogenic service.

Deviation from these normal Code principles may be permitted but only when such deviation is agreed in writing between the Competent Person and the User. The document shall include details of a technical justification and a copy shall be kept in the plant records.

5

EXAMINATION PRACTICES FOR PRESSURE VESSELS

5.1 GENERAL COMMENTS

The types and styles of pressure vessels within the scope of this Code of Practice will be many and varied, but will conform to the definition of the term given at the beginning of this Code. Typical examples are Fractionating Columns, Reactors, Drums, Heat Exchanger Shells and Channels, Receivers and Regenerators.

The vessels may be of simple construction, to relatively simple design Codes or complex both in construction and design. They may have internal linings of refractories or metal cladding of various forms and, externally, may be jacketed and/or painted or lagged and clad.

Therefore the manner and frequency of examination will be conditional on design and construction, materials of manufacture, operating conditions and many other factors. The examination of lined as against unlined vessels requires different procedures and equipment, although in both cases the objective is to establish and conserve the integrity of the pressure shell.

5.2 FACTORS AFFECTING EXAMINATION FREQUENCY

The choice of the examination frequency is allied to the correct Examination Grading Allocation. The principles governing the choice of Grading are given in Chapter 3, with commentaries on Grade Allocation and examination requirements but other factors will also be taken into account in determining the examination frequency, including the following:

(a) severity of service duty and probable deterioration mechanisms,
(b) national legislation where applicable,
(c) normal works overhaul policy,
(d) the ability to carry out meaningful 'on-stream' inspection including monitoring of known acceptable defects and relevant process conditions,
(e) availability of replacement components or materials,
(f) the optimum utilization of existing components and materials,
(g) catalyst life and frequency of regeneration,
(h) quality of internal linings and their resistance to process conditions,
(i) previous history of similar vessels under similar conditions,
(j) remaining design life,
(k) predicted remanent life.

It may also be appropriate to consider consequences of failure.

5.3 CAUTIONARY NOTES

In addition to the factors noted above in Section 5.2, a short commentary is given below on other matters which should be taken into consideration.

(a) Lined Vessels

Linings may be applied internally to protect the pressure shell from thermal, corrosive or erosive conditions. In the event of a lining failure the pressure shell will be exposed to conditions which it was not designed to withstand.

It is the possibility of lining failure and the consequences arising therefrom which must be taken into account when selecting the examination intervals, together with the speed and facility with which lining failures may be detected whilst the vessel is in-service.

(b) Internal Fittings

The presence of internal fittings and the difficulties of removing some of them will be a further factor influencing examination intervals. Parts of the shell may be inaccessible and it is often those areas which may be most susceptible to corrosion due to an inability to clean the areas adequately, creating a trap for a corrosive environment. All internal fittings must be removed as necessary so that the shell can be examined in an adequate manner consistent with the requirements of an examination, unless it is agreed by the Competent Person that the examination can be satisfactorily carried out without their removal.

(c) External Lagging, Cladding and Fireproofing

Sections of the outer surface of the shell may need to be examined at each Examination, as well as holding down bolts, skirt attachments or the surfaces under saddle supports. Consideration should be given to periodically removing sections of lagging or fireproofing, particularly when temperatures are between − 5 and + 140 deg. C for significant periods, in order to expose the outer surfaces to detect the presence and extent of external corrosion.

Serious consideration should be given to examination of vessels constructed of austenitic steels at areas of stress concentrations for possible surface cracking due to chloride ions from atmospheric sources or leached from the insulation and which accumulate and concentrate.

5.4 PREPARATION FOR EXAMINATION

Thorough and adequate examination is dependent on the equipment being prepared in a proper manner. This will necessitate opening up the vessel for internal access, cleaning and the possible removal of some internal components to permit movement through the vessel. Reference may be made to the IP Model Code of Safe Practice, Part 3, Refining Safety Code, for guidance on safety matters in relation to making equipment safe for access and work.

It may be necessary to require the removal of further internal components — to provide access to the vessel

shell internal surfaces which are normally inaccessible — such as tray plates, baffles, weirs, shrouds, wear plates, etc.

Safe and adequate means of access to the vessel, both internally and externally, must be provided. The surfaces and components to be inspected must be adequately cleaned and lighting and other essential services must be available.

Under PSRs in the UK, such preparatory work may need to be included in the 'Scheme of Examination.'

5.5 RECORDS

After each examination, whether in-service or pre-commissioning and after each Grading Review, a Report shall be prepared by the Competent Person(s).

Each such Report shall:

(a) state the extent of the examination,

(b) detail the condition of the equipment including components and/or fittings,

(c) quantify condition found,

(d) state examination methods used,

(e) reference relevant repair weld and NDE procedures,

(f) specify any repairs, removals or modifications required or carried out and any changes to the Safe Operating Limits on future safe life predictions,

(g) review the Scheme of Examination including the examination grading previously selected,

(h) set the date that the next examination falls due.

Examples of Examination Reports are given in the Appendix.

Appropriate persons or Departments responsible for the operation and maintenance of the equipment shall be made aware of the contents of the Report and the Report shall then be added to the Records File for the items concerned (see Section 2.1).

5.6 'ON-STREAM' EXAMINATION METHODS

A number of methods may be used to acquire data and information relative to the condition of the equipment whilst it is in service or relative to the internal environment of the vessel which in turn quantifies the corrosive or erosive nature of the service. The information acquired can be most useful as an adjunct to the normal visual examination, and can play an important part in assessing the condition of a vessel. However in general it will not replace an examination with the vessel off-stream, opened up and cleaned. Due regard must be given to the inherent limitations of this form of examination.

6

EXAMINATION PRACTICES FOR PROTECTIVE DEVICES

6.1 GENERAL COMMENTS

For the purposes of this Code, items of equipment encompassed by the term Protective Devices are described in Definitions, Section 1.3(g). All such Devices should be registered according to the recommendations given in Chapter 2.

Historically the majority of such devices were fitted to protect the vessel from the effects of excess pressure although it was common to fit fusible plugs to the combustion chambers of shell boilers to protect them from the effects of excess temperature. This chapter is sub-divided into two groups: pressure-actuated devices, and other devices. Consideration should be given to including the piping associated with these devices in the scheme of examination, particularly where blockage could occur.

6.2 PRESSURE-ACTUATED DEVICES

6.2.1 Safety Valves

Should be subject to the requirements of Table 1, Chapter 3, and in no case exceed a period of 72 months between examinations.

The valve should be removed from the Process facility and immediately tested on a suitably calibrated Safety Valves test rig to check the pressure at which the valve would have lifted in service. The results of these tests shall be recorded, as subsequent examination intervals may be influenced by the performance history. Following this, the valve should be dismantled, cleaned, repaired, restored, lapped and reset to its correct cold differential test pressure.

6.2.2 Pilot-operated Safety Valves

These devices are generally fitted to low pressure high volume gas systems. They consist of a small pilot relief valve which senses the system pressure and in the event of overpressure causes the main relieving device to open. It is not normally practical to test the main device in a workshop and testing is restricted to the pilot valves.

6.2.3 Bursting Discs

(a) Installation

Bursting discs are protective devices designed to operate once only and cannot be subject to meaningful tests prior to installation or in situ. It is therefore important that special care be taken in their manufacture and assembly and that the correct materials are used.

The device should be inspected at the assembly stage and the disc material selected from guaranteed batches of which samples have been previously tested and for which certificates of material properties have been supplied. The disc should be properly identified and a final thorough check made when the device is installed to ensure that it has been installed correctly.

(b) Replacement

Bursting discs should be renewed at suitable intervals to avoid their premature failure in service due to fatigue or corrosion. It may be useful to conduct bursting tests on discs removed from service to determine whether or not exposure to service conditions has affected their bursting pressure, and this information may be used as a guide to the frequency of renewal.

6.3 OTHER DEVICES

Items included under this heading are as detailed below, but should be included as Registered Protective Devices when they are essential to prevent a dangerous situation from arising.

(a) Fusible plugs. These devices may occasionally still be used on air receivers and steam generating shell boilers but are never fitted to process pressure vessels.

(b) Thermal trips, sensors and alarms. May be tested in the works, in situ, or in the workshop by direct or indirect means, that is, the application of heat, or an e.m.f. equal to that generated by the sensing couple. In addition, all instruments systems should be tested in the field for continuity and, where possible, set point.

(c) Instruments, or devices which are used to monitor or control process variables which, should they deviate outside of predetermined limits, would permit unsafe conditions to be reached, should be examined and tested at intervals appropriate to their duty and electrical/mechanical characteristics.

Examples are high integrity protective devices, corrosion monitors, pH monitors, level glasses and level alarms, mixture ratio controllers and/or alarms.

(d) Non-return valves, which may not be positively depended upon to prevent excess pressure being built up within a pressure vessel from some source at a higher pressure. They may, however, be used to prevent back flow which may:

(i) cause too rapid a loss of pressure which in turn can damage the pressure vessel materials or its metal components, and/or

(ii) allow undesirable fluid to enter the pressure vessel.

They should be examined, tested and/or maintained at intervals appropriate to their duty.

6.4 ISOLATION PRACTICES

Precautions are required to ensure that vessels are adequately protected against overpressure at all times. In particular it should not be possible to isolate pressure relieving devices from any vessels that they are designed to protect unless one of the following provisions applies:

(a) Multiple pressure relief devices. Any provision made for isolating any one relief device for testing or servicing ensures that the remaining relief device connected to the vessel provides the full capacity required.

(b) Single pressure relief devices. Provision is made for their removal for testing or servicing by the use, for example, of an automatic shut-off valve, where the valve is retained in the fully open position by the presence of the relief valve and closes before the relief valve is completely removed. In carrying out such a procedure it is essential that the vessel is not left unprotected and that a replacement relief device is fitted immediately.

(c) Simultaneous isolation of relief and pressure source. The only source of pressure which could lead to an unsafe condition originates from an external source and this source is also isolated from the vessel with the relief device.

6.4.1 Bursting Discs Beneath Safety Valves

Bursting discs are occasionally fitted on either side of safety valves to protect them from the effects of corrosive atmospheres or where absolute freedom from leakage under normal operating conditions is desired. The space between the bursting disc and the safety valve must either be vented in such a way as to positively prevent a pressure rise in the space due to slow leakage through or past the disc which would equalize the pressure across the disc and so prevent its correct operation or fitted with a pressure indicating device and/or alarm. 190

The condition of the bursting disc and the venting arrangements should be checked at the same time that the safety valve is removed for overhaul examination and test.

7

MODIFICATION AND REPAIR

7.1 MODIFICATIONS AND REPAIRS

Any proposed modifications to the design of the pressure vessel or repair to the pressure parts of the vessel should be subject to the prior approval of a Design Authority and the Competent Person. The design and execution of the modification or repair should be carried out under the control and surveillance of a Competent Person, refer, Reg. 9 of PSRs.

When designing modifications or repairs to the pressurized parts of the system, whether temporary or otherwise, consideration should be given to the original design specification, the duty for which the system is to be used after the modification or repair, including any change in relevant fluid, and the effects any such work may have on the integrity of the pressure system, and whether the protective system is still adequate. The repair or modification should be adequate for duty as compared with the original design specification and in accordance with appropriate standards.

Full documentation of the modification should be retained in a Records File. This will include full technical details of materials and techniques, drawings, test certificates, etc. and the original of the approvals procedure documents, containing the signature, position and professional status of all those persons responsible for authorizing, designing, carrying out and testing of the modifications.

7.2 EFFECTS OF MODIFICATION OR REPAIR ON PROTECTIVE DEVICES

Prior to any modification or repair being authorized, full consideration should be given to the possible effects of that modification or repair on the protective device or devices associated with the pressure vessel.

Particular attention should be given to the setting and capacity of pressure relieving devices, both safety valves and bursting discs, and this attention should extend to closed relieving systems downstream of the pressure relieving device and any changes in back pressure that may affect the capacity or setting of the pressure relieving device associated with the pressure vessel under consideration plus the possible effects on the pressure relieving devices fitted to other vessels and discharging into the same closed relieving system.

7.3 STRESS-RELIEVED OR HEAT-TREATED VESSELS

Care must be taken to ensure that vessels which have been stress-relieved or heat-treated as part of their manufacturing process are not subject to welding repairs or hot work of any kind without specific approval for the proposed method and/or techniques being obtained from the Competent Person. The welding procedure, materials, weld set-up, equipment and final heat treatment must also be approved.

8

TESTING

8.1   GENERAL COMMENTS

There are essentially five forms of testing applicable to pressure vessels.

(a)   Strength Testing

Through the application of an applied load, usually greater than the maximum load generated in service but less than that which would cause physical damage to obtain demonstrable proof that the vessel could safely withstand the service load, e.g. hydraulic or pneumatic pressure tests.

(b)   Leak Testing

Through the application of a pressure differential across the pressure vessel to detect leakage paths or leakage rates through the 'pressure envelope' of the vessel. The pressures applied, liquid or gaseous, may be much less than the maximum service pressure, e.g. vacuum tests, search gas tests, air tests and water tests.

(c)   Non-destructive Testing

Through the application of methods based on physical phenomena leading to the detection of imperfections in the materials of construction of the vessel, e.g. radiography, ultrasonics, crack detention, surface replication, etc.

(d)   Destructive Tests

Through the application of loadings to specimens of materials of construction to determine the ultimate mechanical properties of those materials, e.g. yield tests, UTS tests, impact tests, rupture and creep tests, etc.

(e)   Materials Analysis

To establish the chemical or physical characteristics and constituents of materials used in the construction, e.g. spectrograph or chemical analysis.

Any or all of the above tests may be applied at various stages in the life of the pressure vessels. Items (d) and (e) will usually be confined to activities associated with construction, erection, modification or repair. Consideration should be given to protection of adjacent areas during pressure testing.

8.2   PRINCIPLES OF STRENGTH TESTING

The statutory requirements of some countries require periodic strength tests of pressure vessels in order to verify and demonstrate their continued fitness for service. The pressure applied may approach that used in the original Code test, which is itself based on the design pressure, with a suitable correction factor applied for the ambient/design temperature differential, and for deviations from the 'new and cold' condition.

Strength tests should also be applied when repairs or modifications have been carried out which affect the structural integrity of the pressure envelope. Refer also the 'A Guide to the Pressure Testing of In-Service Pressurized Equipment,' EEMUA Publication No. 168 – obtainable from EEMUA, 14–15 Belgrave Square, London, SW1X 8PS.

8.2.1 Hydraulic Testing

The following safety factors must be borne in mind whenever conducting strength tests:

(a) Care must be taken to avoid the possibility of the water freezing within the vessel, which could rupture the shell due to expansion.

(b) The vessel foundations and supporting structure must be checked for their ability to support the combined weight of the vessel and the liquid required to fill it.

(c) The temperature of the material under test must be above that at which a brittle failure may be initiated. In this connection, account must be taken of the lateral shift of the temperature transition curve of certain high strength alloyed materials exposed to high temperatures for prolonged periods.

(d) When filling vessels adequate vents must be provided at all high points, to avoid entrapped air.

(e) When emptying vessels adequate vents must be provided to avoid the possibility of drawing a partial vacuum in the pressure vessel shell.

(f) In determining test pressures, due allowance should be made for the actual condition of the vessel under consideration. Usually, the equipment is not in the 'new and cold' condition or original construction. The test pressure should be gradually applied and, following the test, gradually released.

(g) The specified test pressure includes the amount due to static head at any point under consideration. Particular care should be taken when applying an hydraulic test to a vertically mounted vessel, which may have been tested initially in the horizontal plane.

(h) Care must be taken to ensure that pressure indicating devices used in tests are properly connected and accurately calibrated.

8.2.2 Pneumatic Strength Testing

Pneumatic tests should be avoided wherever possible but, when it is absolutely necessary (as when foundation problems preclude the use of water) the test pressure should be limited to a maximum of 100 per cent of design pressure. Pressure should be applied in small increments, and should be reduced to a pressure not exceeding the design pressure before any close examination is carried out. Only personnel who require to be present for the test should be allowed in the vicinity of the equipment. *Note*: personnel should not be in the vicinity of the vessel when it reaches maximum pressure applied, and this should be reduced before examination for leaks/distortion etc. is carried out.

Pneumatic tests should only be carried out under close supervision and when adequate means have been provided for remote observation. Non-essential personnel should be evacuated from the immediate test area. The likely consequences of a vessel rupture should be fully assessed before starting a pneumatic test.

8.3 PRINCIPLES OF LEAK TESTING

Leak testing is for the purpose of detecting leakage paths through the pressure envelope, in either direction. The methods used may be straightforward water or air pressure tests or by the use of search gases. Leakage detection may be visual or by measurements of pressure drop or by pressure or vacuum decay rate, trace gas response to detection instruments, audible or ultrasonic devices, fluorescent dyes, etc.

Where search gas techniques are used, the pressure applied need be only of a very low order in relation to the working pressure of the vessel under test.

Where audible or visual techniques are used, higher pressures may be required to detect leakage.

APPENDIX

<u>EQUIPMENT INSPECTION REPORT</u>

Block   :⁻ _____   Date of Inspection :⁻ _____

Unit   :⁻ _____   Report No.   :⁻ _____

Equipment No. :⁻ _____   Service   :⁻ _____

Type of Exam. :⁻ _____   Page 1 of ____
==

Next Inspection Due   :⁻ __/__/__   Thorough Inspection Interval :⁻ _____

Type of next Inspection :⁻ _____   Next Thorough Inspection Due :⁻ __/__/__
==

<u>CONTENTS</u>   <u>Page Number</u>

<u>Purpose of the Inspection</u> ....

<u>Preparation</u> ....

<u>Parts not Inspected</u> ....

<u>Results of the Inspection</u> ....

<u>Details of Work</u> ....

<u>Recommendations</u> ....

<u>Attachments</u> ....

Examined by :⁻_____ (Print)_____ (Sign)_____ (Date)

Approved by :⁻_____ (Print)_____ (Sign)_____ (Date)

	REPORT OF EXAMINATION OF VESSEL (CLASS B)	Form INS/3	Sheet 1 of 2
		Reg. number	
		Local identification	

PART 1 - DESCRIPTION

Title

Works	Plant	Unit
Date of Examination	Date of last Examination	Inspection class, category and grade
Date of Construction	Has last report been scrutinised?	

If equipment has a steam jacket see additional report (Form F58) dated

Brief History

PART 2 - VESSEL DATA
(Copy relevant data from equipment file)

	SHELL	TUBE SIDE	JACKET/COIL
Operating Fluid			
Design Pressure(s)			
Design Temperature(s)			
Thickness			
Nominal Corrosion Allowance			
Protected by Relief Stream N^{os}			

PART 3 - EXTENT AND METHOD OF EXAMINATION

What examination and tests were made?

What parts were not examined?

PART 4 - SUMMARY REPORT

	MATERIAL	INTERNALLY	EXTERNALLY
Shell			
End(s)			
Tube Plate			
Tubes			
Branches			
Welds			
Vessel Supports			

DEFINITION OF TERMS

G:	GOOD:	SATISFACTORY:	POOR:
As new without sign of corrosion.	Surface rust slight scale or corrosion not exceeding 0.5 mm (20 Thou) and within the corrosion allowance.	Corrosion greater than 0.5 mm (20 Thou) and within the corrosion allowance.	Heavy corrosion and/or deep pitting or cracking of any description. Reassess inspection frequency. IA: Inaccessible NA: Not applicable

PART 5 - DETAILED REPORT

PART 6 - REPORT OF PRESSURE TESTING (Where applicable)

COMPONENT	SHELL	TUBE SIDE	JACKET/COIL
Test Medium			
Test Pressure			
Test Period			
Date of above test			

Observations

PART 7 - STATEMENT BY INSPECTOR

The vessel is suitable for its design duties specified in Part 2 subject to the relief system remaining in order .... Yes/No

Does the Scheme of Examination require amending as a result of this examination? .... Yes/No

I certify that this document is a true report of the examination specified in Part 3 and carried out on the ....

The interval to the next thorough examination has been reviewed and has been retained/changed to ....

Inspection Grade............................ The next thorough examination in due before................... ....

The next intermediate examination in due before............... ....
(if appropriate)

Computer Coding of Next Examination (if appropriate) ....

Signed .... Date ....

Endorsed (where applicable) by .... Date ....

PART 8 - STATEMENT BY WORKS ENGINEER AND/OR NOMINEE

The above equipment may be used within the duty specified in Part 2, subject to the above conditions

Signed .... (Works Engineer
and/or Nominee)

Date(s) ....